the Fruittella roll

marcel verhaaf

publisher's details

BIS Publishers
Building Het Sieraad
Postjesweg 1
1057 DT Amsterdam
The Netherlands

T +31 (0)20 515 02 30
F +31 (0)20 515 02 39

bis@bispublishers.nl
www.bispublishers.nl

ISBN 978-90-6369-303-9

Design: Marcel Verhaaf
and Paulien Hassink
www.phontwerp.nl

www.iconicpackaging.com

contents

iconic packaging 4 is the Fruittella roll an icon? 6 80 years of Fruittella sweets 9 the secret recipe for the Tella dough 23 how the Tella dough is made 24 some facts and figures 29 packing Fruittellas 30 the packaging through the years 36 Fruittella goes global 45 Blacktella 48 advertising 52 TV advertising 54 the slogan 56 SQuare Games 58 Fruittella spreads its wings 61 acknowledgements and sources 64

iconic packaging
marcel verhaaf

Consciously or subconsciously, in our daily lives we put icons on a pedestal. In most cases, we deem them more important than anything that isn't iconic. Why is that? And how do we decide what constitutes an icon?

Opinions differ as to what an icon actually is. My preferred definition is as follows: an icon is a visual representation which makes the underlying relevance immediately apparent to a wide group of people. By this token something is an icon if, upon seeing it, a large group of people instantly think of its deeper meaning.

People can be icons – Winston Churchill and Nelson Mandela being two examples. Upon seeing pictures of them, most of us are immediately aware of their social relevance. Equally, the logos of brands that are directly associated with a particularly target group, lifestyle and so on can also be icons. Harley Davidson is a prime example. Even simple products or packaging can be icons.

If there is no natural relationship between the image and its meaning, we use the term symbol, rather than icon. In this case, the meaning is derived from an agreement that the users of the symbol have made. In this sense, the majority of road signs and national flags are symbols. We have mutually agreed, for instance,

that a round sign with a horizontal white stripe means "No Vehicular Access". Because there are so many shades of grey in between, I won't linger too long over the definitions. I will leave that to experts in semiotics, who have published plenty on the subject.

My fascination with icons relates primarily to brands and packaging. A number of these are without doubt true icons, as their representation instantly calls to mind a deeper meaning. After the end of World War Two, for example, for many Europeans the Lucky Strike cigarette packs became synonymous with liberation. They were brought home by the allied soldiers and often distributed to the population of the liberated countries. Most packaging icons have a less heroic backstory. Rather, they owe their deeper meaning to an association with a particular target group, life philosophy, innovative trait or positive nostalgic value.

When we are talking about packaging, I feel a distinction must be made between "icon" and "iconic". In design spheres, if packaging is described as iconic, this usually means that it has certain traits that make it unique, striking and resonant, and easily understood by the target group. History teaches us that the most iconic packaging does have a unique appearance at the time of its launch, which sets it apart from the norm. It is also clear and simple.

That said, an unusual, simple and recognisable design does not, in itself, make a product's packaging an instant icon. The criteria of "deeper meaning" and "familiarity to a wide public" must first be met. Various mass media can help achieve this. Often, a product's packaging has a small surface area and is rarely big enough to adequately communicate its deeper meaning.

The more frequently we encounter a product, the greater the likelihood that we will notice it
Another important aspect of "iconisation" is the repetition factor. The more frequently we encounter a product, the greater the likelihood that we will notice it and become interested in it. And the greater the likelihood that we will be persuaded that it is relevant to us, and genuinely worth considering as a purchase.

On the face of it, this is a fully conscious process - but unfortunately that isn't really true. Even more important than our conscious thought processes is the role that our brain plays in purchase decisions - as a number of recent neurological studies have proved. Importantly, by brain we do not mean our conscious thought processes. Our brain seems to favour familiarity over unfamiliarity. As neuroscientists tell us, this is the consequence of the evolution of the human brain over the last 40,000 years. The brain is hardwired to distinguish between things that are potentially dangerous and things that are safe. Between things it recognises and things it does not recognise. In other words, between the familiar and the unfamiliar.

All of which suggests that repetition helps. Along with a distinctive and simple design that is easy to recognise and remember - and means it can easily be stored by the brain. Iconic design is the prelude to creating a true icon.

The Heinz ketchup bottle is a great example. Its appearance has been almost unchanged for 120 years: a bold and simple design that is familiar to many millions of people around the world. One neurological study found that, when an image of the famous Heinz ketchup bottle was shown to respondents, they experienced activity in the same part of the brain that reacts when respondents are faced with images of their friends...
A remarkable fact, to say the least.

Does this mean that iconic packaging can become a true friend?

is the Fruittella roll an icon?

Let's get one thing straight: in the history of design the Fruittella roll has not achieved a status comparable to that of the Coca Cola bottle or the Campbell's soup tin. There can be no doubt that these are true design icons in the eyes of many. They have earned their place in museums and original examples are highly prized by collectors.

I've yet to see a Fruittella roll in a museum and there are hardly legions of collectors of Fruittella-related memorabilia. And yet, I believe the Fruittella roll does have true iconic value. My aim, in publishing this book, is to raise awareness of its iconic aspects. Better still, by devoting an entire book to the subject and increasing the repetition factor, I am seeking to further the iconic cause of the square Fruittella roll. A little outside help can go a long way. After all, Andy Warhol was primarily responsible for elevating a number of supermarket products to iconic status. Thanks to him, they became figureheads of the pop-art movement.

We use the term icon when the design is a visual representation of a deeper meaning. Moreover, that deeper meaning must be immediately recognisable and relevant to a large group of people. Some examples of this are the Harley Davidson logo or an image of Winston Churchill with his perpetual cigar. The former is a representation of a particular rebel lifestyle; the latter symbolises the allied victory in 1945. Both, therefore, are a unique, distinctive and recognisable image with a deeper meaning.

The "back seat sweet"

What can be said about the deeper meaning of Fruittella? To many consumers, Fruittella is nothing more, nor less than a delicious sweet. For a large group of Dutch consumers in the 45+ age range, however, the Fruittella roll does have a special significance. Back when they were children, foreign holidays were rapidly gaining in popularity. Lots of families headed south by car - suitcases teetering on the luggage rack and the kids in the back seat. Fruittella was an essential travelling companion, a welcome sweet treat on a tense yet exciting journey. Let's not forget that, in those days, foreign travel was still a new and thrilling experience for many families. For them, Fruittella still represents that feeling - so it is something of an icon after all.

In common parlance, iconic is a term that refers chiefly to an object's appearance. In the case of packaging, it suggests that the design is distinctive and unique. That it "stands out from the crowd". The Fruittella roll more than fulfils that description.

Fruittella-reclame op een van de bedrijfs-auto's.

Strictly speaking, it is not a roll: it can't roll, for starters. Which is why the people who make the sweets at Van Melle prefer to call it a "bar". In the world of confectionery, it is still an unusual shape. Most other sweets are packaged in bags and round rolls. But the main drawback of those round rolls is that they do just that: they roll. And before you know it, they are back-to-front. That's not an issue for Fruittella. Obviously, the Fruittella roll owes its shape primarily to the individual, square sweets it contains. In the factory, they are called "chews".

As you will see in this book, the graphic design of the packaging underwent a number of changes

in the early years. Until recently, therefore, it was chiefly the shape of the packaging, rather than the whole entity including the printed design, that was considered iconic.

In recent decades the printed design has, in essence, remained unchanged: a solid red logo on a background that indicates the flavour, followed by a small illustration of the fruit in question. Every few years, the design is tweaked to ensure that the image is always in keeping with the times, but the overall design remains recognisable.

Nowadays, the name Fruittella is a generic name, a product name and a brand name. Remarkable!

Fruittella advertising on one of the company's vehicles.

MADE IN HOLL...

*Newly-weds
Izaak and Adriana
van Melle (1900)*

80 years of Fruittella sweets

In 2012, the company Perfetti Van Melle celebrates the 80th anniversary of the journey of Pierre and Machiel van Melle. They brought the original Tella dough to the Netherlands - a defining moment in the history of the Fruittella brand.

What better excuse to look back at the factors that made the Van Melle business successful. One of the key factors is undoubtedly the pioneering mentality of the founders. Over the company's more than 112 years of existence, rising prosperity has been another factor instrumental in enabling more and more people to afford luxury confectionery items. Demand increased at an incredible pace. It wasn't just the growth in disposable income that brought confectionery within the reach of a wide public: equally important was the steadily falling price of the key ingredient, sugar.

Sugar as an ingredient

The history of confectionery can be traced back to the ancient Egyptians. 2000 years BC, they satisfied their sweet cravings with combinations of fruit, nuts and honey. They had yet to discover the use of sugar as an ingredient. The first written record of sugar as a solid substance dates back to 500 BC, in India. The consumption of sugar across swathes of the globe was not possible until sugarcane began to be cultivated on a wider scale. Until well into the 19th century, the Europeans remained dependent on expensive imports of raw cane sugar from the "East" for their sugar needs. The colonisation of South and North America changed all that. Sugar plantations were established, where sugar could be produced much more cheaply. In 1747, German scientist Andreas Marggraf discovered that sugar could also be obtained from beet. Thanks primarily to French efforts, in the 19th century ever-greater quantities of sugar were produced from beet. Until that time, there was no sugar production in Europe. The increasingly large-scale production of sugar as a result of the Industrial Revolution made sugar accessible to almost everyone.

Sugar processing

With the availability of sugar came the craft of sugar processing. In the Southern Netherlands, there are references to the "expert preparation" of confectionery way back in the early 16th century. Hardly surprising, given that the port of Antwerp played an important role in importing cane sugar. The small area between the Scheldekade and the Grote Markt was home to around a hundred "confectioners". In the 17th century, Antwerp grew into the world's largest sugar processing centre. In his bakery, which he had taken

over from his father in 1882, Abraham van Melle employed a young Belgian lad, who brought with him the confectionary preparation skills he had learned in the south. The Belgian "sugar balls" the bakery had begun to produce were a huge hit and increasingly large quantities were delivered to individuals and shops. As time went on, the bakery began making boiled sweets and importing "drops" and "biscuits" from England. In 1897, the popularity of these products is illustrated by the fact that 10,313 kilos of confectionery were sent in a single ship.

After working for a long time in his father's business, Izaak van Melle decided to realise his big dream and set up in business himself, preparing confectionery. His father-in-law had pledged the necessary funds, enabling him to open his new shop on 9 October 1900. The shop was located opposite Abraham van Melle's premises and, reportedly, was also fitted out with some equipment from his shop. The fact that this was not a bitter rivalry but a close relationship is clear from the fact that Izaak was also "gifted" two customers when he started up his business. In the early years, Izaak supplied his ice cream bonbons, silk violets, Viennese bonbons and other delicacies mainly to buyers in the Zeeland region and a handful of other places, namely The Hague, Vught and Kampen.

Izaak's professionalism and passion for quality extended to the packaging of his products. In 1902, the first 6000 labels were ordered for the "Bonbons Suisse". The following year, proper company labels were produced, sporting an eagle as the trademark. This was an essential move in consolidating the company's reputation. In those days, the quality of many foods was very variable. A guarantee of origin from a famous producer was an important recommendation - to the extent that some competitors had the audacity to use the actual name "Van Melle".

Most products were supplied in bottles and tins, and their buyers sold them to end consumers by weight or even "individually". Aside from the ice cream bonbons, virtually no confectionery products were individually wrapped. To prevent their contents sticking together, it was very important that the bottles and tins had airtight seals. Izaak had a tough time getting his suppliers to comply with this.

His involvement in the packaging was not confined to the technical aspects. He personally decided what would appear on the increasingly colourful labels. The Morocco fudge bonbons were one example: "The label must

depict a "Moroccan in National costume, presenting the sweets within, with 3 half-moons in the blue sky and, alongside, Morocco cough sweets, written in clear letters. Beneath this somewhere, the words "reliable cough remedy"." He was an art director avant la lettre. Van Melle recognised early on the importance of eye-catching packaging. He also realised that, in those days, the way in which the packaging was decorated and covered with attractive images was a means of persuading shopkeepers and consumers of the product's quality. After all, the products of the Zeeuwsche Dropsfabriek were being delivered further and further afield - even to places where not all the consumers were aware of the company's reputation. In the absence of mass media, it was the vendor behind the counter and the packaging that persuaded consumers of a product's merits.

In 1907, the first decorated tin cans were supplied

In 1906, the tin printing company Haarlemsche Blikdrukkerij and Dordrecht-based metal goods factory Wed. J. Bekkers en Zn. were approached to discuss the possibility of decorating the actual tin cans. One year later, the first decorated cans hit the shelves, depicting wintry scenes for the ice cream bonbons and four different fruit designs for the "kussentje" sweets.

On 10 February 1912, the company became a public limited company, which was to continue its life as Van Melle's Fabriek van Suikerwerken and, in its international trade, Van Melle's Confectionary Works.

By around 1913, the scope of its business activities had widened considerably. By that time, the company was operating an electricity power station and an "Electric Laundry, Bleaching and Ironing Establishment".

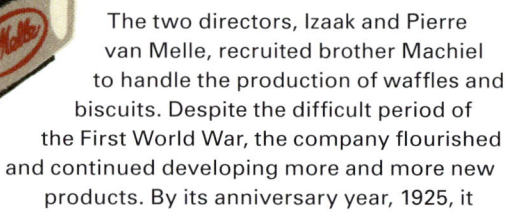

The two directors, Izaak and Pierre van Melle, recruited brother Machiel to handle the production of waffles and biscuits. Despite the difficult period of the First World War, the company flourished and continued developing more and more new products. By its anniversary year, 1925, it was exporting to nearly 100 ports around the world and generating turnover of 1.18 million guilders. In a separate department, the fruit from its own fields and orchards was cleaned, cooked and preserved.

In 1926, a new product was added to the already huge range: the toffee. Izaak and Pierre had discovered this product in England. Among the flavours they produced were vanilla toffees, chocolate toffees, nougat toffees and the white "polar bear" acid sweets. The toffees were individually machine-wrapped in greaseproof paper and delivered to stores in magnificently printed cans. They were real works of art - particularly when you bear in mind that lithography was used to apply the decoration to the tin. The mechanised half-tone print process had yet to be invented, which meant that 10 to 15 colours were regularly required in order to make the images – most of which had a natural theme – look as attractive as possible. The cans became more and more elaborate, with designs as eclectic as doctor's bags and complete mills with revolving sails. Almost a hundred different tins were marketed. It almost beggars belief that hundreds of thousands of these luxury cans were produced. This was the heyday of printed tin in the Netherlands. In the 1930s, the shop-sold tins got competition in the form of magnificent counter-top jars, which had the advantage of making the – often beautifully wrapped - sweets within more visible to customers.

At the end of the 1920s, the first signs of the impending crisis became apparent. To keep their machines operating, businesses scoured the globe in search of new opportunities and new products. Pierre and Machiel van Melle set off for Poland, after learning that the company Lax und Sohn had developed a unique procedure for manufacturing fruit chews. After some negotiation, Van Melle acquired the manufacturing rights. Some time later, these chews were to arrive on the market as Fruittella.

In 1935, the range comprised 90 varieties of biscuit, 50 varieties of waffle and 176 varieties of drops

The range reached a new high point around 1935. There were 90 varieties of biscuit, 50 varieties of waffle and 176 varieties of drops. To serve customers even faster and, of course, in order to make a statement, the Netherland's first ever commercial company aeroplane was acquired and was aptly named the "Double Eagle". Customers could even save up for a round trip, by collecting 24 of the vouchers contained in each tin of Monospar mix.

The war years proved disastrous for Van Melle – as, indeed, for almost every company in the Netherlands. Not just from an economic perspective, but also in a personal sense. Some of the company's male employees were even deported to Germany in 1944. In that same year, Breskens suffered repeated bombardment by English planes trying to head off the retreat of German troops. Van Melle's Confectionery Works was completely destroyed. After the liberation, the company was keen to get up and running as soon as possible. Because it would take too long to rebuild the site in Breskens, the confectionery works were relocated to Rotterdam. In the interim period, toffees and biscuits were produced in Vlaardingen, until a completely new factory was opened in Rotterdam. This was officially inaugurated on 22 November 1950, by the mayor of Rotterdam. That year, most of the confectionery was packaged by machine. For the biscuits and waffles, it would take a few more years before they could also be packaged by machine. At the end of the 1950s, the huge range had to be reduced, as it was becoming far too expensive to produce and maintain stocks of all the different varieties. Not that business was still bad: in fact, sales were buoyant in the years following the reconstruction.

It wasn't just in the Netherlands that things were slowly improving. There was a renewed upturn in demand from abroad and, with its new factory in Rotterdam, the company was able to respond better than ever to that increased demand. Foreign sales trips also proved successful. A sales office was opened in Antwerp, a retail organisation was established in Germany and, from 1956 onwards, the company even began production in Brazil, albeit on a modest scale. Oddly enough, the biggest stumbling block to international growth was the staff shortage in Rotterdam. To tackle this problem, the company relied increasingly on mechanisation. At the start of the 1960s, Fruittella and Mentos became the heroes of the range. Mentos really took off once the company began selling these "little discs" in rolls rather than three kilo boxes. International markets were becoming increasingly important. By 1962, half of the company's total output was exported. Amidst the social upheaval of the 1960s, the company enjoyed steady and comfortable growth.

It was not until the 1970s that the company embarked upon a radical change of course, placing even greater emphasis on developing its foreign markets. This policy created a 25/75 split between the domestic and foreign markets in terms of share of sales. A number of years later, this ratio was to become 15/85. Quality improvements, product development and more market-led adjustments remained part and parcel of the company's philosophy. In contrast to confectionery products, biscuit production had long been in the doldrums. In fact, at the start of the 1970s, this business was loss-making. Although manufacture made an important contribution to covering general overheads, the decision was made to gradually run that side of the business down. By the end of the decade, the company was producing biscuits chiefly for the German market.

The 1970s also heralded a very different approach to communicating with consumers. By then, self-service had become the norm, and this had far-reaching implications for premium brands. It had taken a while for this American style of shopping to catch on with Dutch consumers. The Netherlands' first self-service store was opened in 1948, but the model didn't really take off until the 1960s. By 1968, the share of the supermarkets had risen to 72%. This brought about the demise of many grocers who, until then, had dispensed advice and recommended certain branded products from behind their counters. With the self-service stores came an increasing number of own-brand products, which competed with the products of the famous brands. To persuade consumers of their qualities, these famous brands had to start using mass media on a large scale.

... if it's toffee, it has to be Van Melle. Van Melle Toffees, the tastiest ever!

....toffees moeten van *van Melle* zijn

van Melle

TOFFEES, HEERLIJKER DAN OOIT!

Van Melle was no exception. The first advertising short for Van Melle was broadcast on Dutch television on 29 June 1976. The first in a series of advertising and promotional campaigns to strengthen the Mentos and Fruittella brands, it was followed by promotional films in cinemas and a packaging facelift. Thanks also to more efficient production methods, in 1977 the group increased its turnover by an impressive thirteen percent. Further growth was to come between 1980 and 1983, when turnover doubled and profit increased by 50 percent. This was a formidable achievement, but something of a double-edged sword, as. there was no further scope to increase capacity at the factory. Once again, the decision was made to build a new factory. Meaning another drastic relocation.

The Enka complex in Breda-Noord became vacant and proved the most suitable location. The new factory officially opened on 28 September 1984. The 1980s also brought a major change in company's financial structure. Van Melle "went public". This initial public offering paved the way for significant future expansion and, in 1991, enabled Italian company Perfetti, another family-owned business, to acquire a 37 percent share.

The strong ties with the Italian market date back to 1973, when an agreement was concluded with Perugina for the sale of Van Melle confectionery products. Perfetti subsequently took over the sales activities and, thanks in part to successful television advertising, was to make Italy Van Melle's biggest export market.

Perfetti began life as 'Dolcifcio Lombardo' in 1946, near Milan, where brothers Ambrogio and Egidio Perfetti initially produced sweets and bonbons. The brothers' decision to start producing chewing gum marked a huge turning point. American soldiers had brought this new product with them to Italy at the end of the Second World War. Although it was the first chewing gum to be produced entirely in Italy, the brand name "Brooklyn" is an unmistakeable nod to its American origins.

In Italy, the brand was also known as "la gomma del ponte". The formation of the company "Gum Base", which produced the basic ingredient of chewing gum, proved a very shrewd strategic move. By introducing a large number of successful brands, which were backed up by progressive advertising, the company became a powerhouse in the international confectionery sector. A series of takeovers of large companies, among them the Belgian company Frisk, were to follow in the 1980s and 90s.

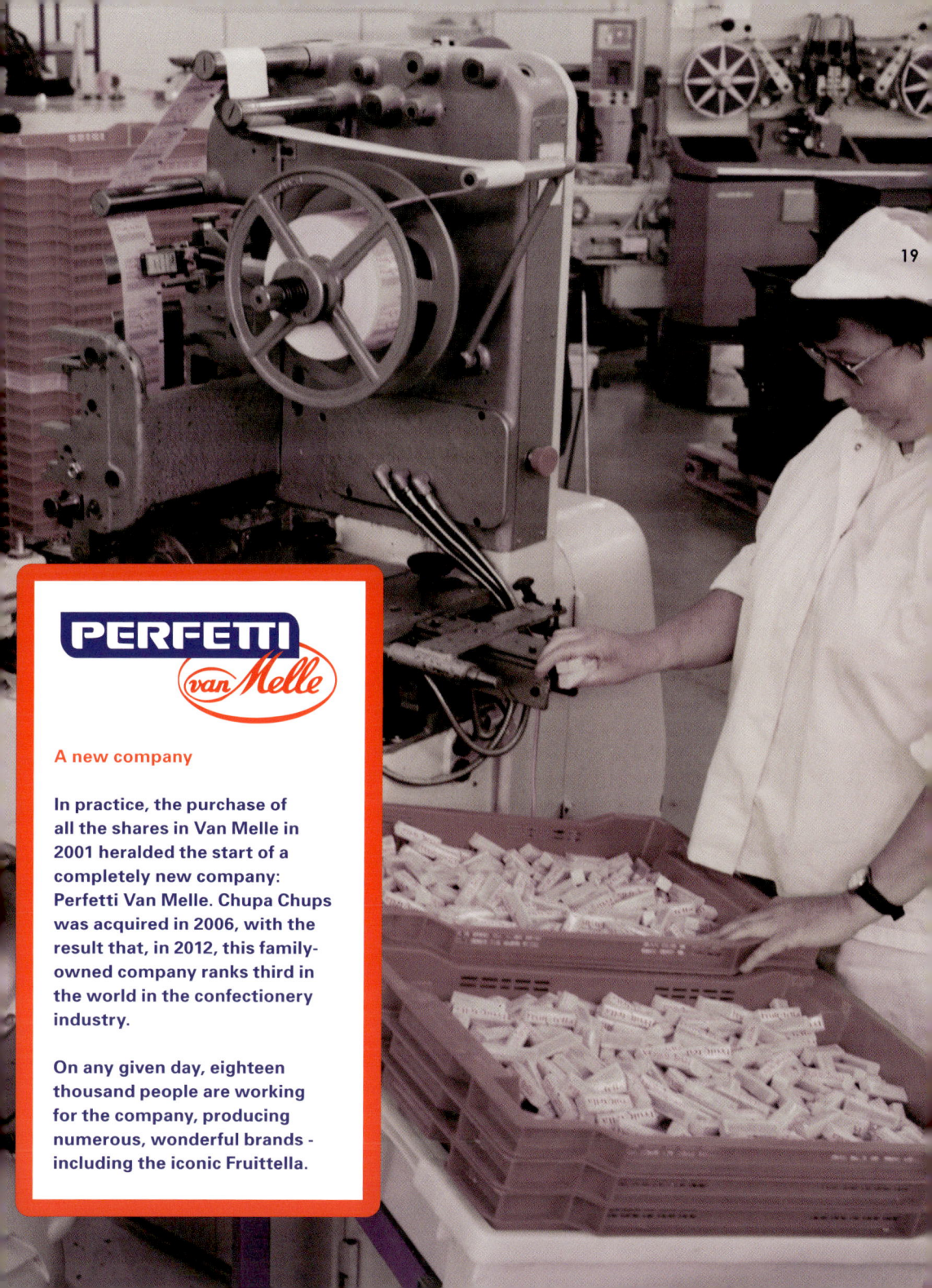

A new company

In practice, the purchase of all the shares in Van Melle in 2001 heralded the start of a completely new company: Perfetti Van Melle. Chupa Chups was acquired in 2006, with the result that, in 2012, this family-owned company ranks third in the world in the confectionery industry.

On any given day, eighteen thousand people are working for the company, producing numerous, wonderful brands - including the iconic Fruittella.

Tella's toffees
Delectably soft

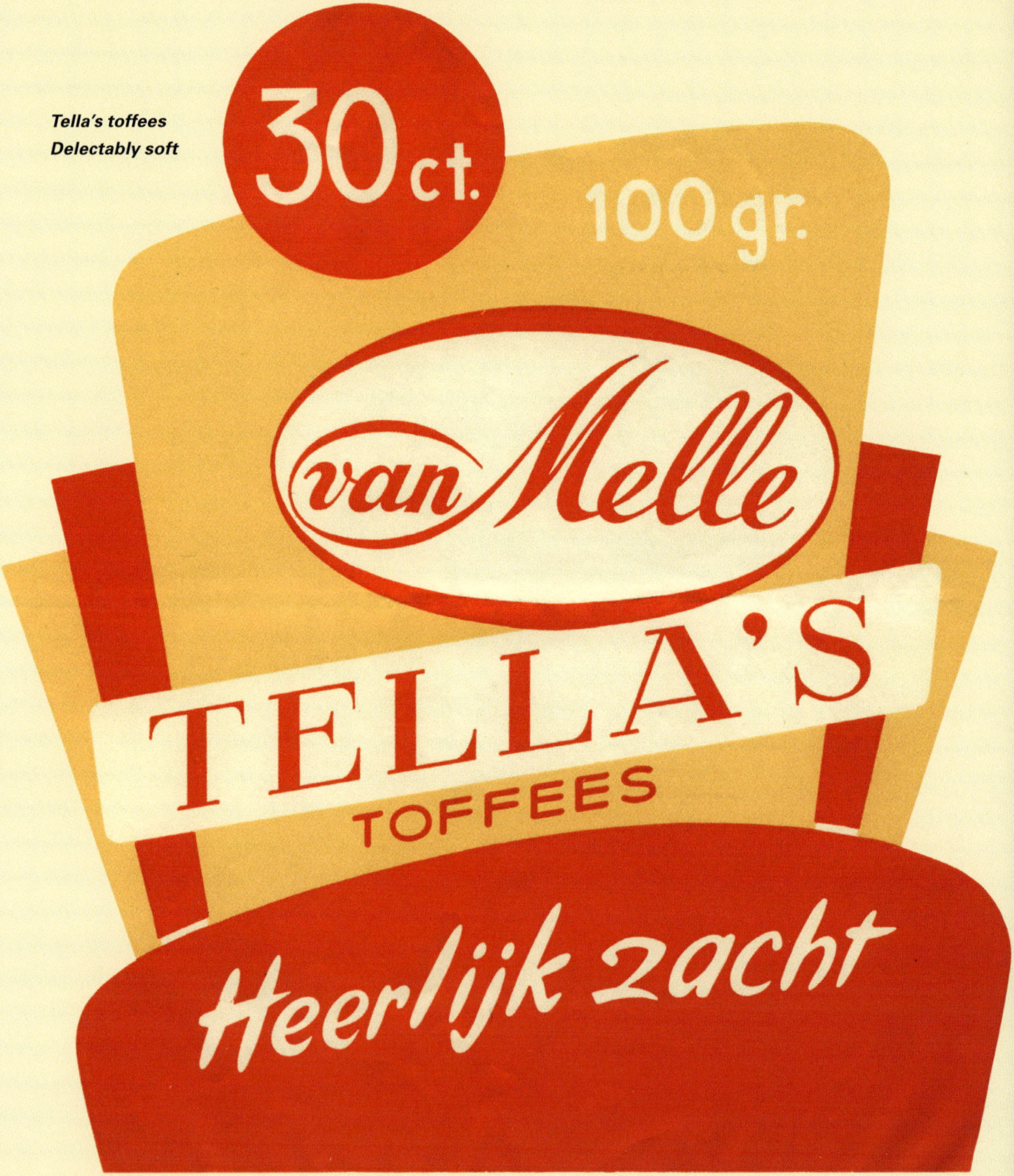

the secret recipe for
the Tella dough

At the start of the 1930s, Van Melle had heard about a Polish company that had developed a very interesting method for preparing soft fruit chews. The success of this recipe was clear from the fact that various companies in Switzerland, Austria, Germany, France and Belgium had already bought the rights. In May 1932, Pierre and Machiel van Melle paid a visit to the confectionery company Jozef Laz und Sohn in Krakow, where they observed the secret procedure for making soft fruit chews and, eventually, negotiated the signing of the contracts.

Those contracts related to the rights to sell the product in the Netherlands and Dutch East Indies. Exports to other countries would be penalised by a 5000 guilder fine. Violations of manufacturing secrecy would also entail severe penalties. That secrecy had to be guaranteed for a period of 6 years, after which manufacturing rights would pass entirely to Van Melle. Therefore, in the first few years the recipe was a closely-guarded secret. For a long time, it was Pierre van Melle himself who assembled the dough ingredients. Aside from him, almost nobody knew the combination of the ingredients and binding agents.

> **"The soft chews didn't end up tasting of soap and didn't get damp if they were wrapped in the prescribed packaging"**

In return for abiding by these strict rules, the company acquired the means to create a product that had lots of potential. According to Lax, the benefits were obvious: "The soft chews didn't end up tasting of soap after a few months and didn't get damp if they were wrapped in the prescribed packaging. They wouldn't go off, providing they had been prepared according to the recipe and stored in the manner that is customary in the confectionery trade."

These fruit chews, which were a novelty in the Netherlands, were wrapped individually in paper and marketed under the name Fruittella. Fruittella really took off after 1950, when a completely new form of packaging, the "packet", was introduced, to replace jars, tins and boxes.

It was the birth of our iconic square roll. In the ensuing years, the introduction of the packet time and again provided a rapid sales boost on the various export markets.

how the Tella dough is made

The right chewing consistency and the ingredients are what make Fruittellas so delicious

There are two very important aspects involved in making Fruittellas so delicious. First, of course, the ingredients, which determine the different flavours. The second and, without doubt, the most important factor is the perfect chewing consistency, which all hinges on the method used to prepare the dough.

The entire process begins with combining ingredients including sugar and glucose syrup. The sugar crystals are dissolved by adding water and bringing the mixture to the boil. The composition of this mixture is very precise. The mixture must always be made according to the correct steps.

Next, fruit juices and natural flavourings are added. All the ingredients are thoroughly mixed in a kneading machine and, at the same time, the dough is cooled.

This cooling process creates new, tiny sugar crystals. If the kneading is done correctly and the temperature and humidity are exactly right, the famous Fruittella structure is created. This is an 80-year-old secret! After kneading, the dough is shaped into a long slab for further processing.

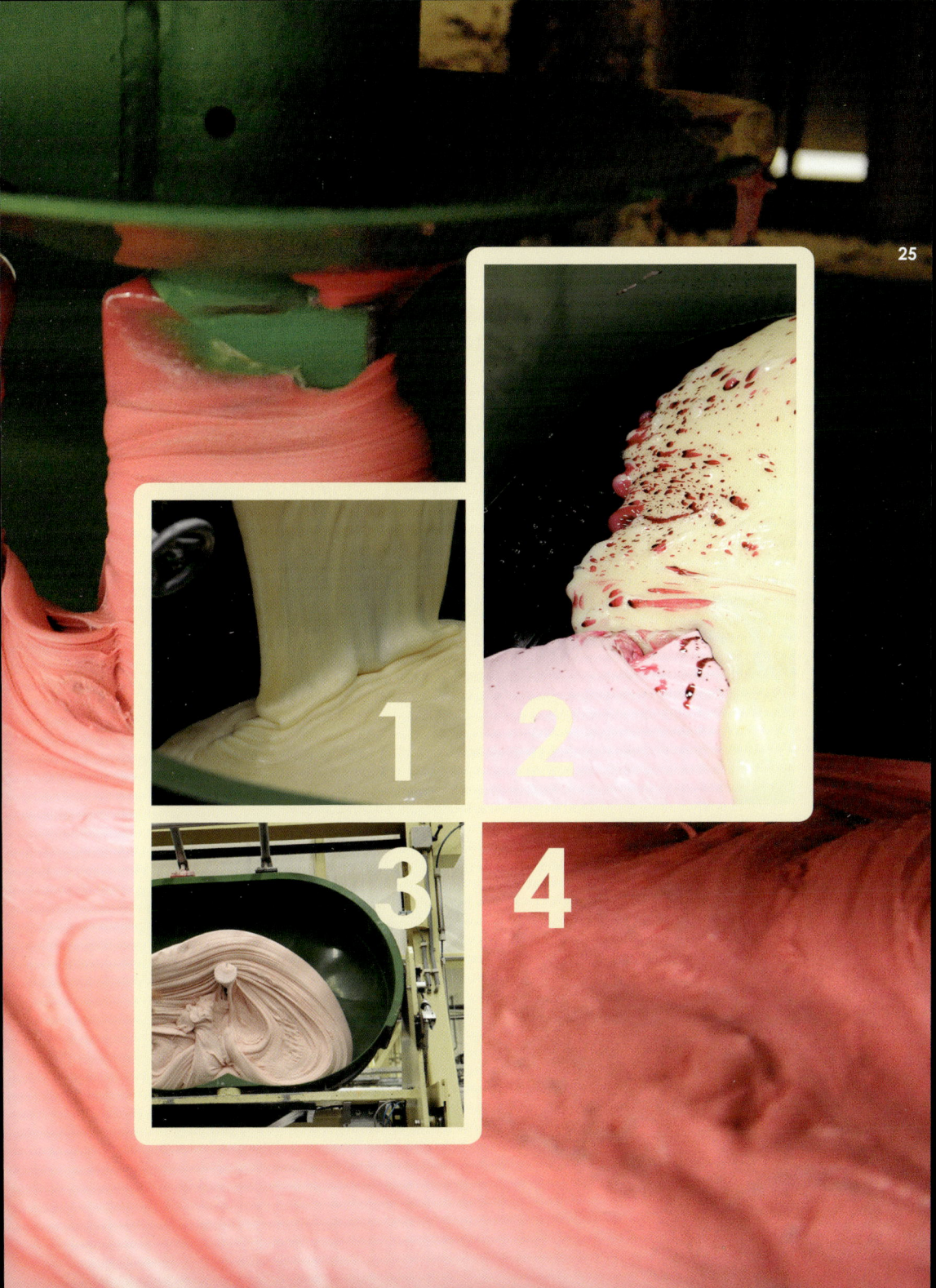

how the Tella dough is made

Maintaining a consistent level of quality is a tricky business

For the final time, the shape of the dough changes and the next machine in the line transforms it into a long, round "snake". The round snake is then shaped into a square snake. This is the point at which the typical Fruittella ridged effect is created on the top of the chew. Small chunks are then cut from the snake, and all that remains is to wrap them. The Tellas are ready to eat.

It sounds simple but in reality, of course, maintaining a consistent level of quality is a tricky business. The cooking temperature and the percentage of humidity in the dough are influenced by the outside temperature, atmospheric humidity and even the barometer reading. Nowadays, cooling tunnels and air conditioning help keep the entire process under tight control.

In the old days, it was almost all done by hand. Even the process of getting the taste and colour just right was intuitive. This required a great deal of experience, because certain flavourings and fruit juices react differently to others in the dough mix.

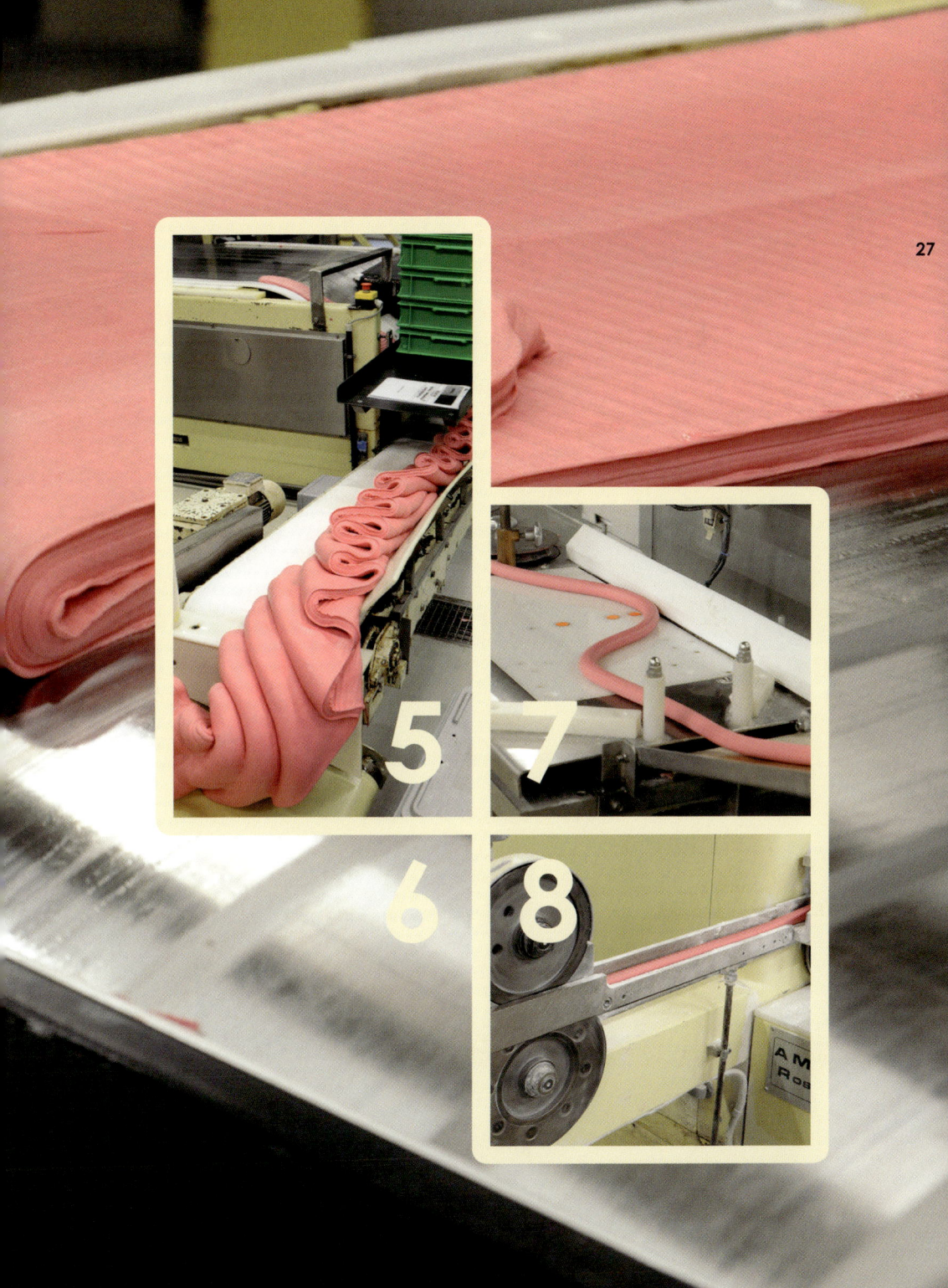

5

6

7

8

Squarely the fruitiest.

When all the Fruittella
machines are
operating simultaneously,
500,000 Fruittellas are
produced each hour.

***

If you laid all the Fruittellas
produced in one week
in a row, they would
stretch 1020 kilometres.
That's the distance from
Breda to Milan.

***

It takes one-and-a-half hours
from sugar grain to
Fruittella.

***

Fruittella is produced
in 6 countries.

***

After 25 years,
an employee of the Fruittella
production plant will have
made 45 million litres of
Tella dough.

packing Fruittellas

The packing process begins with the "cut & wrap machine". A square strand of Tella dough is fed into one side of the machine, which first cuts off small chunks of dough. The wrappers are then cut from a roll, to exactly the right length, and wrapped around the chunks.

Because it happens so fast, this process takes place in a sealed section of the machine. An operator attends the machine, to check quality and ensure that the various flavours of Tella dough are placed in the correct wrappers.

2 **3**

Fruittella chews are caught in plastic crates as they exit the machine. The fresh chews are still soft and very fragile. They are briefly left to cool in the crates, which makes them firmer and ensures that they retain their square shape as they undergo the subsequent stages in the packing process.

packing Fruittellas

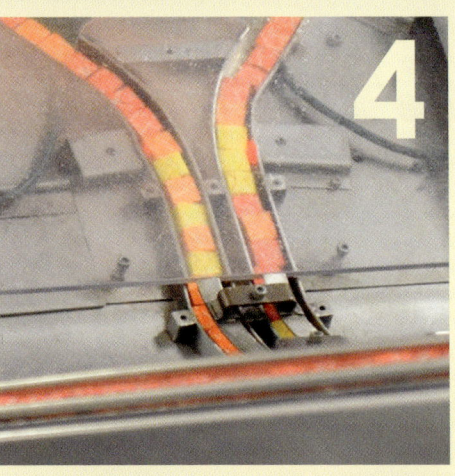

4

Each roll of Fruittella contains ten chews. A roll can contain a single flavour or a combination of flavours. Just before the outer wrapper is applied, a machine checks that there are ten chews. If not, they are ejected via a side exit. They can then "join the back of the queue" again, to be given a second chance.

The outer wrappers are applied at high speed, as an infinite roll of printed paper. Little black squares are also printed on the images of the wrappers, called "cutoff spots". These tell the machine where the wrappers start and end, so that they can be cut at exactly the right point.

5 **6**

packing Fruittellas

7

The Fruittella rolls are then packed in boxes. In the past, a lot of people were needed to neatly pack all the rolls in the counter display boxes. Modern, fast machines have now replaced humans, and make it possible to continue production in the Netherlands. Fortunately, people are still needed to thoroughly check the quality of the chews.

8

the packaging
through the years

The very first Fruittella rolls date back to the early 1950s and were a real novelty in stores. Until that point, most sweets had been sold loose. Van Melle called them "packets".

The packets were silver, printed with a continuous pattern of individual fruits and yellow stripes with the words "Van Melle", "FRUIT" and "TELLA'S". The advantage of this pattern, which was also known as "scattered printing", was that it eliminated the need for absolute precision when cutting the individual wrappers. Although effective, the printed design soon had to be adapted to the changing requirements created by the emergence of self-service stores. With the arrival of these stores, packaging had to tempt consumers to make a purchase. The recommendation of the grocer behind the counter would become less and less relevant.

1954

17ct

't Nieuwste
van 't
Nieuwste

1956

the packaging
through the years

White packaging with a pattern of thin silver lines

Fruittella was repackaged in the late 1950s. The new packs were predominantly white, with a pattern of thin silver lines in the background. As before, the eye was drawn to the Fruittella logo and a realistic illustration of a fruit.

Although modest in size, the Van Melle logo had also acquired a permanent place above the Fruittella logo. The person behind this design had come up with another clever idea. The rolls were too narrow to accommodate a "proper" image of fruit. So, the illustration was wrapped around the corner, onto the next side, on which a Fruittella logo was also placed. When the rolls were arranged alternately in the counter display boxes, the fruit illustrations matched up on different sides, revealing the entire fruit. The effect was far more attractive than a small illustration.

Because of this, the packing department at the factory had to be particularly careful when packing the boxes. The white packaging remained in use until the 1970s.

Buy two, get one free! Another new treat from Van Melle.

1957

1959

Fruit-tella

1963

the packaging
through the years

**Typical
1970s design
is also in
evidence
on rolls of
Fruittella**

In the 1970s, marketing people became increasingly interested in the role of packaging as a communication tool. Packaging that, in some cases, had remained virtually unchanged for decades was adapted to the needs of the day. The flowery, graphic forms so inextricably linked with the 1970s found their way onto rolls of Fruittella.

Because not all the markets had embraced the latest design trends, increasingly a hotchpotch of packaging began to emerge until, in the early 1980s, the decision was made to introduce uniform packaging around the world, and select a single design. It was just as simple and effective as the old, white packaging, but much more colourful. Each fruit flavour was given its own background colour. And the new logo featured, in the largest possible font: a combination of words and a picture of fruit.

1975

1988

41

2001

the packaging
through the years

In 2002 a design was chosen which no longer incorporated both the typography and the fruit illustration. The Fruittella word mark was supported by a green shape beneath, creating a uniform visual cue for the logo. One year later, the green shape had already been abandoned in favour of a pictorial mark on a background colour that changed depending on the flavour. To emphasize that only natural colours and flavourings are used, an illustration with a more natural appearance was used.

A packet containing five different flavours

The very latest packaging is for Fruittella RAINBOW, a packet containing five different flavours. The flavours of the chews within correspond exactly to the colour codes and fruit illustrations on the exterior. This might seem perfectly straightforward to consumers. But for the people in the factory, getting everything just right takes a lot of time and concentration.

2002

2007

2012

VAN MELLE
MINT
TELLA'S
MENTHOL EUCALYPTUS
van Melle
Euca-tella
MENTHOL EUCALYPTUS
van Melle
Mint-tella

Fruittella goes global

Since the very early days, Van Melle as a company has been characterised by its focus on growth and, later, international expansion. Whilst the two world wars and the crisis of the 1930s were major setbacks, the van Melles have always bounced back.

The first "proper" exports began in 1904, to the Dutch East Indies. Prior to that, the company had already begun selling to the Belgian market. Given the geographical location of its headquarters in Zeelandic Flanders, this wasn't really considered to be abroad. The company soon extended its export activities to the British East Indies, Algeria, Morocco, Greece, Asia-Minor and Syria. In the ensuing period, more and more export destinations were added. And the company even set up foreign production sites, including in Brazil in the 1950s.

Along with many other varieties of confectionery in the range, Fruittella became increasingly popular in a growing number of export markets. Much to the amazement of Gideon van Melle and his father Izaak, Fruittella also proved a big hit in England - the very country where, in the early years, the company had gained a lot of its knowledge. The introduction of the packets provided a further boost to sales. And, from 1958, sales were further helped by advertising in the underground stations with the slogan "Hey Fellas, Fruittellas".

Van MELLE MINT

The launch of Fruittella in Japan is an example of the major changes that have to be made to accommodate specific markets.

Of the original seven flavours that were considered for the Japanese market, two were decreed suitable to launch on a trial basis.
However, the chosen varieties turned out not to be quite what consumers wanted, necessitating adjustments to the chewing consistency and flavours. Moreover, the name "Fruittella" was not, it was discovered, suitable as a brand name. Apparently, it had negative associations in Japanese. After conducting a survey of 600 respondents, the name "Fruiteen" was chosen.

More so than in other parts of the world, "the packaging" has great symbolic value to the Japanese. In Japan, it is customary for people to give each other small gifts, but the care taken over the packaging is deemed more important than the contents. Hardly surprising, then, that the Japanese have high expectations of the packaging on offer in supermarkets.

For Van Melle, the Japanese launch in the 1970s presented a challenge to the company's passion for quality. Special care was taken over the choice of paper, the outer boxes and even the adhesive tape used to seal the boxes. Under Japanese law, the packaging must not adhere to the product. Furthermore,

there were no end of strict requirements with regard to the composition of the inks used. The Japanese climate fluctuates between extremes of very cold and hot and humid weather, which means the packaging must be insulating and watertight.

In the end, the company had to introduce additional production stages, in order to create packaging that met the requirements in terms of the right appearance and the technical prerequisites of the packaging machines. The result was a high-gloss material, which involved printing first on a layer of film before the aluminium layer, the paper and a hot melt wax were applied. In those days, this was quite a technical feat.

Fruittella goes global

Strawberry flavour was popular everywhere in the world and was produced in four varieties

As the company began exporting to more and more countries, demand arose for products that better reflected consumers' differing flavour preferences. Strawberry flavour, which was popular everywhere in the world, was therefore produced in four varieties. A liquorice version was added to the range: "Droptella", and a toffee flavour with the rather splendid name "Caratella de Fruittella". This latter variety was far more expensive to produce. It contained more ingredients than the fruit-flavour chews and required a much longer cooking time.

The flavour inventors at Van Melle developed many more exciting products, including grapefruit, extra sour, marzipan, tropical (peach, mandarin and passion fruit) and Milk Chews. The yoghurt/strawberry and blueberry/yoghurt flavours were developed especially for the Russian market.

There were even found to be big differences around the world in terms of preferred chewing consistency. Most Europeans, for instance, want to be able to chew first, before the product dissolves gradually as you chew.

The Americans prefer a large piece of candy that dissolves quickly, whilst many people in Asia have a preference for buying products with a long chewing time.

Blacktella

A "love it or hate it" flavour that is only popular in a few countries

One unusual product within the Fruittella family is the liquorice-flavoured candy chew. In the Netherlands, these chews were launched under the name "Droptella". It's a "love it or hate it" flavour that is only popular in a few countries of the world. In those countries, however, this product has a large and loyal following. In Germany, where an impressive two million packets are sold each year, the Droptellas are marketed as "Lakritz Toffee". A fruit variety is now available in Germany, 'Lakritz Frucht-Mix Toffee', which makes the close family ties with Fruittella even more apparent.

Blacktellas were originally developed for the Italian market and had a slightly different taste, thanks to the addition of a touch of menthol. This very distinctive flavour went hand in hand with a rather unusual aroma. Employees from the factory who were working on the original Blacktella recipe claimed that, at the end of the day, they themselves smelled of Blacktella. Despite the differences in flavours, from around 1977 the same simple and effective packaging design was rolled out to every country. It was all black, with the product name emblazoned in large silver letters and the Van Melle logo in red. Except for a few details, this design has remained unchanged to this day and still ensures that the Lakritz Toffee stands out among the many other sweets and snacks in the shop.

E CANDY CHEWS
S TENDRES À LA RÉGLISSE
RTIFICIEL
LLE MORBIDE ALLA
A
:

8710 8316

tella®

tella®

GRATIS rijtoer met ponywagen

Gouden koets…

…inlevering van een ledige **verpakking:**

VAN MELLE
Bakkertjes
of

VAN MELLE
Clubwafel
of

VAN MELLE
London Fair
of

VAN MELLE
Café Noir
of

Fruit-tella
of

rol Mentos

is hier

FREE pony & trap or Golden coach ride… on presentation of an empty pack of:
Van Melle Bakkertjes, or Van Melle Clubwafel, or Van Melle London Fair, or Van Melle
Café Noir, or Fruittella or Mentos roll. The Van Melle Circus is here…

advertising

Advertising has a huge influence on establishing the iconic status of packaging. I would go as far as to say that good advertising and promotion are vital. In order to be iconic, a product's packaging must be effective and clear. And distinctly different to everything else in its category. Sometimes, the only decoration on the packaging is a logo. Alternatively, a company might choose a symbol that says more about the brand's mentality than about the characteristics of the product within.

Good, iconic packaging is the representation of the underlying message, rather than the message itself. Many famous examples of iconic packaging originate from the period before the Second World War, when they were seen primarily as a means of identifying and protecting the product.

The majority of the information that the manufacturers wanted to convey to consumers was shared through advertising or the store owner. The brands used mass media to communicate their message. For decades, packaging was "left alone" and, thanks to advertising, the designs remained simple, effective and often beautiful. In recent years, manufacturers have had

to operate according to much tighter budgets. They have less money for advertising and have to share the money that is available among more and more different media. As a result, packaging must increasingly assume the role of advertising. All the information about the brand, the product and its use must be conveyed chiefly by the packaging - with the result that packaging is becoming complicated, cluttered and, increasingly, downright

Increasingly, packaging has to assume the role of advertising

unattractive. When advertising can be used to explain what the brand stands for and what the product means to us, the design of the packaging can be striking, effective and unique. Fruittella is a good example. Since the very beginning, the design of the rolls has been clear and simple.

Fruittella's status as the "travelling product par excellence" has been communicated since the very first advertisements from the 1950s: "Fruittella – the natural travelling companion." These were no idle words. Because of its unique chewing consistency, a Fruittella chew stays in the mouth for a long time.

Fruittella the natural travelling companion...

There's no more refreshing company than the little square sweet, in five fresh fruit flavours! And...

They're just so good to chew!

TV advertising

In the first ever TV advert for Fruittella, we see a boy with his mother at the dentist. What makes this short so unusual is that the dentist is explaining "Fruittella's kind to teeth", because "it doesn't stick fast." A dentist promoting sweets is a message that would be inconceivable these days. The boy in the advert is none other than Marius van Melle, from the van Melle family. This film already focuses on the square shape of the sweet, which provides the backdrop to the entire film.

*Fruittella
doesn't stick,
so it's kind to teeth.*

*It's true:
it really doesn't stick
to your teeth.*

Fruittella by Van Melle is the most refreshing sweet around. That's why it's my Number One choice.

Let me tell you something!

the slogan

The slogan "De enige rol die niet kan rollen van de smaak" (which roughly translates as "The only roll that doesn't cut corners with flavour") pithily conveys what, nowadays, is the most iconic aspect of the Fruittella packaging: a square roll. Quite a bold move, considering that most rolls of sweets are indeed round. It all starts, of course, with the product itself, which is square.

The reason for Fruittella's square shape is explained in a series of commercials from the 1980s - short cartoons in which a professor explains, among other things, why a square sweet is packed with more flavour than a round sweet. The cartoons were drawn in the studio of Marten Toonder, who was also responsible for the Tom Puss and Oliver B. Bumble comic books that are famous in the Netherlands.

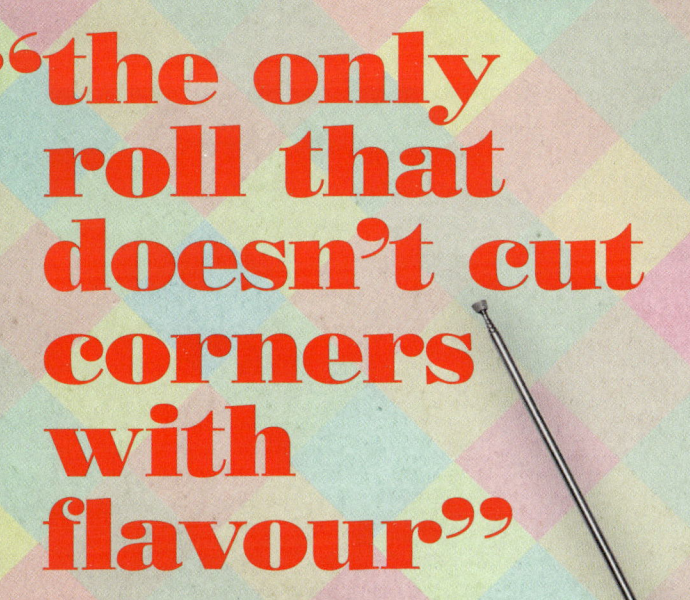

"the only roll that doesn't cut corners with flavour"

SQuare Games

There are many different features that can make a product's packaging iconic, such as its shape, brand image or unique colour. The distinguishing feature of Fruittella, of course, is its "squareness". The chews are square, and so is the roll.

The SQuare Games were invented in the Netherlands to force home the brand's iconic status among consumers. The competition involves a number of well-known games which instantly become even more challenging when they are square.

In 2011, the SQuare Games were broadcast on Dutch television, in the form of short mini-competitions.

Just imagine: SQuasketball with square balls and square crates, or SQuala Hoops with square hoops.

And, of course, people could join in and play SQuare Games online…

Fruittella spreads its wings

With its 80-year history, Fruittella is now a household name in the confectionery market - thanks primarily to the success of the square chews and the square rolls. But there's much more to Fruittella than just chews.

Obviously, the unique Fruittella dough can be used to make all kinds of delicious treats. The flavour inventors at Van Melle are endlessly creative.

It started with the Fruittella sticks, which appeared in shops back in the 1950s. These were the first products aside from the square chews to actually bear the Fruittella name. Initially, Fruittella was a product name that was used alongside the Van Melle brand. It was common to ask: "would you like a Fruittella?" As the years went by, the name Fruittella became more and more familiar and, eventually, somewhat eclipsed the name Van Melle in the minds of consumers.

So, from 2000 onwards, the company decided to sell all the other products made using Tella dough as a base under the Fruittella brand. Nowadays, the Tella family is a large one, comprising Stix, Crunchies, Minis, Pixels, Dummies and Lange Jans.

With all those different shapes, they certainly won't fit in square rolls. That privilege is still reserved for the original chews.

Russia, Jellies In Russia, Fruittella is the market leader for sweets. Not just with the square packet, but also with bags of chewy sweets in the shape of bottles, snakes and worms.

Netherlands, Fruittella Cream Light

A creamy, sugar-free variety, for sucking instead of chewing.

England, Fruit Filled

Filled Fruittellas with a liquid centre.

Brazil, Berry and Jelly

Fruity sweets in a tin.

Brazil, Fruittella Fruit & Choco

The fruity flavour of Fruittella with a smooth chocolate centre.

NEW

Netherlands, Fruittella Snoeppotten *Fruittella introduces a range of sweet pots, filled with fruity, winegum-type sweets. The familiar square shape is echoed here in the design of the packaging.*

63

England, UFOs
Another Fruittella innovation: Round Fruittella sweets with a crunchy shell and delicious, fruity flavour.

England, Pixels
Small, crunchy Fruittella sweets in long, narrow bag.

Netherlands and Belgium, Fruittella portion-pack range
Portion-packs of Dummy, Mini, Lange Jan, Crunchies and Pixels. This portion-pack range is specifically for children; the products are also tailored entirely to children, in terms of flavour, texture, packaging and portion size.

acknowledgements

A big thank you to Simon van Dijk, Product Manager Fruittella at Perfetti Van Melle Benelux. Simon was the driving force behind this project on behalf of Fruittella. Without his input, it would have been impossible to obtain all the information needed.

Harald Engelen also deserves a special mention in these acknowledgements. He guided us through the production process at the factory in both the physical and virtual sense.

sources

Werken met suiker (Working with sugar)
Van Melle (1900-1985)
Bram Oosterwijk, Van Melle, 1985

Sweet Memories
A selection of confectionery Delights
Robert Opie, 1988

Bonte blikken (Colourful tins)
Tin manufacture in the Netherlands 1800-1990
Drents Museum Assen, 1991

Van Melle Nederland B.V.
Annual reports 1983, 1985, 1987

Photo of tins on page 11:
*Brandnew Design Collectie
@ ReclameArsenaal
www.reclamearsenaal.nl*